The Physics and Technology of Diagnostic Ultrasound: Study Guide

Second Edition

Robert Gill, PhD

THE PHYSICS AND TECHNOLOGY OF DIAGNOSTIC
ULTRASOUND: STUDY GUIDE (SECOND EDITION)

ISBN: 9780987292193 (pbk.)

Every effort has been made in preparing this book to provide
accurate and up-to-date information which is in accord with
accepted standards and practice at the time of publication.
Nevertheless the author, editors and publisher can make no
warranties that the information contained herein is totally free from
error, not least because clinical standards are constantly changing
through research and regulation. The author, editors and publisher
therefore disclaim all liability for direct or consequential damages
resulting from the use of the material contained in this book.
Readers are strongly advised to pay careful attention to information
provided by the manufacturer of any drugs or equipment that they
plan to use.

A catalogue record for this
work is available from the
National Library of Australia

NATIONAL
LIBRARY
OF AUSTRALIA

*Cover photograph: Ripple in still lake on a misty day, Strathgordon,
Tasmania, ID 13137459 © Heidi Koenig | Dreamstime.com*

Contents

A: Introduction

This Study Guide is a companion to my textbook (see below for details). It consists of a series of short questions and brief model answers to these questions. It should be useful to ultrasound physics students and their teachers and clinical supervisors.

Students: You can use this guide to lead you through your studies, helping to ensure that you cover all the required topics adequately. You can also use it to test your understanding later, for example when preparing for tests and exams.

Lecturers, teachers, tutors: This guide should be helpful as a check that you have covered the material appropriately in your technical and clinical teaching. Please note that the questions may resemble written exam questions but they are mostly too broad to be directly useful as such. However, they can be used as a starting point for more carefully worded quiz and exam questions.

Other resources

1. The companion textbook *The Physics and Technology of Diagnostic Ultrasound: A Practitioner's Guide (Second Edition)* (High Frequency Publishing, ISBN: 9780987292186 paperback, 9780648869603 ebook).[1]

2. Free resources available on the companion e-learning website.[2] These include:

    *   the ***Whiteboard Collection***, video presentations with voice-over showing how to solve all of the exercises in the textbook which involve calculation;

    *   *Test your Knowledge*, a selection of multiple choice questions, calculation-based exercises (with answers) and two crossword puzzles to test your knowledge of terminology.

---

[1]*For more details, see https://ultrasoundbook.net*
[2]*See https://sonophys.com.au*

3. *Introduction to Diagnostic Ultrasound Technology*, an 8 - 10 hour online course which covers the fundamentals of ultrasound physics. Accredited by the Australasian Society for Ultrasound in Medicine (ASUM) as suitable for the Certificate in Clinician Performed Ultrasound (CCPU) and Certificate in Allied Health Performed Ultrasound (CAHPU) candidates.

4. *Mathematics for Ultrasound Physics*, a short 4 - 6 hour preparatory course for students who don't have a strong mathematics and science background. It can be very difficult to understand the physics and technology of ultrasound without an adequate knowledge of mathematics. This course covers the essential concepts in more depth than the brief review provided in the textbook.

5. *Artifacts in Diagnostic Medical Ultrasound: Volume 1 Grayscale Artifacts*, by Martin Necas (High Frequency Publishing, ISBN: 9780987292162 paperback, 9780987292179 ebook)

Robert Gill
Sydney, 2020

B: Questions

Chapter 2: Ultrasound interaction with tissue

1. Define the meaning of each of the terms *energy, power* and *intensity*. Name the units generally used for each of these. Which is the most relevant parameter when the exposure of the patient's tissues to ultrasound is being considered?

2. Define what is meant by the terms *wavelength, period* and *pulse duration*.

3. Calculate the wavelength for ultrasound frequencies of 4.5 MHz and 12.5 MHz.

4. Define the meaning of the term *amplitude*. How is the ultrasound intensity related to the amplitude?

5. List four common causes of attenuation of ultrasound in tissues. Which of these is the dominant factor in most circumstances?

6. Calculate how much a 3.5 MHz transmit pulse would be attenuated in passing through 20 cm depth of average soft tissue (assume the attenuation coefficient is $\alpha = 0.5$ dB/cm/MHz). Express your answer in decibels.

7. The ultrasound intensity is reduced by a factor of 29,700 by attenuation. Express this attenuation in decibels.

8. Explain what is meant by the term *depth of penetration*. Describe how the depth of penetration relates to the ultrasound frequency.

9. If the depth of penetration of a given machine is 8 cm when it is operating at 10 MHz, what will the depth of penetration be if the frequency is changed to 4 MHz?

10. Explain what is meant by the term *reflection*. With the aid of a diagram, describe how the beam direction is affected by reflection. Explain why perpendicular incidence gives the strongest echo.

11. An ultrasound pulse passes through an interface between two tissues at perpendicular incidence. The acoustic impedance is 1.55×10^6 Rayls in the first tissue and 1.50×10^6 Rayls in the second tissue. Calculate what fraction of the incident ultrasound power is reflected by the interface.

12. If the reflection coefficient of a tissue interface is 0.015 calculate the percentage of the incident ultrasound power that is transmitted through the interface.

13. Explain the difference between *reflection* and *scattering*. Describe the appearances produced in the image by each of these processes.

14. Explain the mechanism that causes speckle in the ultrasound image. Name two techniques that the machine can use to reduce speckle.

15. With the aid of a diagram, explain what is meant by the term *refraction* and how the transmitted beam direction can be calculated.

16. An ultrasound pulse passes through an interface between two tissues. The propagation speed is 1470 m/s in the first tissue and 1550 m/s in the second tissue. If the incident angle is 35° calculate the transmitted angle. Does this interface have a critical angle?

17. Explain what is meant by the term *critical angle* and why it is significant clinically.

Chapter 3: Pulsed ultrasound and imaging

1. Explain how the ultrasound machine determines where to place each echo in the image.

2. Calculate the depth of a reflector which produces an echo that arrives at the probe 230μs after the emission of the transmit pulse.

3. Name the machine parameter that determines the axial (depth) resolution of an ultrasound image. Explain how the axial resolution can be calculated.

4. Define what is meant by the term *pulse repetition frequency* (PRF). Explain the relationship between the maximum allowable PRF and the depth of penetration.

5. Calculate the maximum PRF that the machine can use if the depth of penetration is 18 cm. Now repeat the calculation for a situation where the depth of penetration is just 4 cm.

6. Explain how the imaging frame rate is related to the depth of penetration and the number of transmit pulses that are required to produce each image.

7. If the depth of penetration is 10 cm and the machine needs 200 transmit pulses to create each image, calculate what the maximum possible frame rate will be.

8. Explain briefly how the beam is scanned when imaging with each of: a linear array probe, a curved array probe and a phased array probe.

9. Sketch the field of view of each of the standard probe types: linear array, curved array, phased array.

10. Name a typical clinical application area where each of the following probes is used: linear array, curved array, phased array.

11. Define with the aid of diagrams what is meant by each of the following terms: *B-mode, M-mode, A-mode*.

Chapter 4: Transducers and focussing

1. Explain the meaning of the term *piezoelectric*.

2. Explain how the ultrasound transducer converts an electrical transmit pulse into an ultrasound pulse travelling into the patient's tissues.

3. Explain how the ultrasound transducer converts ultrasound echoes into electrical signals.

4. Draw a cross-sectional diagram of a probe showing each of the following: transducer, backing layer, matching layer, lens, gel, electronics. Briefly summarise the purpose of each of these.

5. Explain how the width of the ultrasound beam causes a point reflector or scatterer in the tissues to be displayed as a line.

6. Define the meaning of the term *spatial resolution*. Name the machine parameters that determine the spatial resolution.

7. Explain the meaning of the term *diffraction limit* and its significance. Name the machine parameters that affect the diffraction limit.

8. Calculate the diffraction limit (i.e. the angle of divergence θ) for a transducer with an aperture of 2.5 cm operating at 4.5 MHz.

9. Calculate the beam width at focus for a transducer with an aperture of 1 cm operating at 5 MHz focussed at a depth of 4 cm.

10. Explain why standard linear and curved array transducers cannot focus as well in the slice thickness dimension (also known as the elevation plane). Explain the technique currently used in ultrasound machines to focus the slice thickness of these probes.

11. Define the meaning of the term *matrix array transducer*. Explain the two main advantages associated with the use of matrix array transducers.

12. Explain with the aid of diagrams the principle of electronic focussing.

13. Explain the meaning of the term *dynamic focussing*, and why the machine can only use it when receiving echoes, not when it is transmitting.

14. Explain how multiple transmit focussing operates. Name the principal advantage and disadvantage of multiple focussing.

15. Name four modes of operation where the beam must be steered away from the normal direction when a linear array transducer is being used.

Chapter 5: Ultrasound instrumentation

1. Name three user controls that interact with the transmitter, beam-formers and amplifier.

2. Explain the meaning of the term *time gain compensation* (TGC) and why the machine requires this function.

3. Define the meaning of the term *dynamic range*. Explain why ultrasound machines require dynamic range compression. Describe the main ways the image will change when the dynamic range is (a) increased to the maximum and (b) decreased to the minimum.

4. Explain the purpose of the scan converter.

5. Define the meaning of the term *pre-processing*. Name three examples of pre-processing functions and briefly describe what they do.

6. Define the meaning of the term *post-processing*. Name three examples of post-processing functions and briefly describe what they do.

7. Explain the purpose of the cine-loop function.

Chapter 6: Image artifacts

1. Define the meaning of the term *artifact*.

2. Explain with the aid of diagrams the reason why shadowing occurs. Name one clinical situation where it may occur. Indicate whether it is a potentially useful artifact, and if it is useful when it may be of value. If it is potentially a problem, explain how it can be minimised.

3. Explain with the aid of diagrams the reason why enhancement occurs. Name one clinical situation where it may occur. Indicate whether it is a potentially useful artifact, and if it is useful when it may be of value. If it is potentially a problem, explain how it can be minimised.

4. Explain with the aid of diagrams the reason why edge shadowing occurs. Name one clinical situation where it may occur. Indicate whether it is a potentially useful artifact, and if it is useful when it may be of value. If it is potentially a problem, explain how it can be minimised.

5. Explain with the aid of diagrams the reason why propagation speed artifact occurs. Name one clinical situation where it may occur. Indicate whether it is a potentially useful artifact, and if it is useful when it may be of value. If it is potentially a problem, explain how it can be minimised.

6. Explain with the aid of diagrams the reason why reverberation occurs. Name one clinical situation where it may occur. Indicate whether it is a potentially useful artifact, and if it is useful when it may be of value. If it is potentially a problem, explain how it can be minimised.

7. Explain with the aid of diagrams the reason why ring-down artifact occurs. Name one clinical situation where it may occur. Indicate whether it is a potentially useful artifact, and if it is useful when it may be of value. If it is potentially a problem, explain how it can be minimised.

8. Explain with the aid of diagrams the reason why comet-tail artifact occurs. Name one clinical situation where it may occur. Indicate whether it is a potentially useful artifact, and if it is useful when it may be of value. If it is potentially a problem,

explain how it can be minimised.

9. Explain with the aid of diagrams the reason why range ambiguity artifact occurs. Name one clinical situation where it may occur. Indicate whether it is a potentially useful artifact, and if it is useful when it may be of value. If it is potentially a problem, explain how it can be minimised.

10. Explain with the aid of diagrams the reason why beam width artifact occurs. Name one clinical situation where it may occur. Indicate whether it is a potentially useful artifact, and if it is useful when it may be of value. If it is potentially a problem, explain how it can be minimised.

11. Explain with the aid of diagrams the reason why sidelobe artifact occurs. Name one clinical situation where it may occur. Indicate whether it is a potentially useful artifact, and if it is useful when it may be of value. If it is potentially a problem, explain how it can be minimised.

12. Explain with the aid of diagrams the reason why slice thickness artifact occurs. Name one clinical situation where it may occur. Indicate whether it is a potentially useful artifact, and if it is useful when it may be of value. If it is potentially a problem, explain how it can be minimised.

13. Explain with the aid of diagrams the reason why speckle occurs. If it is potentially a problem, explain how it can be minimised.

14. Explain with the aid of diagrams the reason why refraction artifact occurs. Name one clinical situation where it may occur. Indicate whether it is a potentially useful artifact, and if it is useful when it may be of value. If it is potentially a problem, explain how it can be minimised.

15. Explain with the aid of diagrams the reason why mirror image artifact occurs. Name one clinical situation where it may occur. Indicate whether it is a potentially useful artifact, and if it is useful when it may be of value. If it is potentially a problem, explain how it can be minimised.

16. Explain why artifacts are undesirable even when they are not obvious in the image.

Chapter 7: Doppler ultrasound

1. Explain the Doppler principle and briefly outline how it is used in Doppler ultrasound.

2. Explain why the Doppler shift depends on both the blood velocity and the Doppler angle.

3. Explain why Doppler angles greater than 60° should not be used when blood velocity values will be measured from the Doppler signal.

4. Explain why the patient's exposure to ultrasound is generally higher in pulsed Doppler than in colour Doppler or grey scale imaging alone.

5. Explain with the aid of diagrams the operating principle used in continuous wave Doppler. Indicate how the sample volume is defined.

6. Draw a typical Doppler spectral display. Indicate the units for the horizontal and vertical axes, and the significance of positive (above the zero baseline) and negative (below the baseline) values.

7. Define the meaning of the term *Nyquist Limit*. Explain why it is important in Doppler ultrasound. Explain the concept of baseline shifting.

8. Describe with the aid of diagrams the principle of operation of pulsed Doppler. Explain the purpose of the following: sample volume, angle correction, range gate, wall filter.

9. Draw diagrams showing clinical situations where the Doppler system will display (a) flow towards the probe, (b) flow away from the probe.

10. Briefly explain why the pulsed Doppler PRF changes when the velocity scale control is adjusted.

11. Explain the principle of operation of colour Doppler. Draw a colour Doppler display acquired by a linear array probe and indicate the colour box, the colour bar, the tissue (or grey scale) threshold, the beam direction for imaging beams and the beam direction for the colour Doppler.

12. Explain how the machine calculates the colour value to be

displayed at a given point in the colour Doppler image and how it determines whether colour or grey scale information should be shown at that point.

13. List three ways in which colour Doppler is inferior to pulsed Doppler.

14. List the three colour Doppler parameters that the ultrasound machine calculates for each sample volume and briefly describe their role in producing the image.

15. Explain the meaning of the term *variance* in colour Doppler and explain why the variance can be useful in some clinical situations.

16. Explain why the dependence of the Doppler shift on the Doppler angle can make it difficult to interpret colour Doppler images.

17. Define the term *power mode colour Doppler* (or simply power Doppler). Explain why it can be preferred to standard colour Doppler in some clinical situations. Indicate its advantages and limitations relative to standard colour Doppler.

Chapter 8: Doppler artifacts

1. Define the meaning of the term *frequency aliasing* in the pulsed Doppler spectral display. Draw a diagram to show how it affects the spectral display. List five actions that may be taken to reduce or eliminate aliasing.

2. Explain what is meant by the term *high PRF Doppler*. Indicate when it may be useful.

3. Define the meaning of the term *intrinsic spectral broadening*. Explain why it occurs and how it degrades the quality of the information acquired from the Doppler signal.

4. Define what is meant by the term *spectral mirror artifact*. Explain briefly when and why it occurs.

5. Define the meaning of the term *wall thump*

6. Explain why frequency aliasing occurs in colour Doppler and list three actions that can be taken to reduce it.

7. Describe what is meant by the term *twinkling artifact*. Name a clinical situation where it may be seen.

Chapter 9: Haemodynamic concepts

1. Define the meaning of the term *flow resistance* in relation to blood flow.

2. Describe how variations in the vascular structure cause different organs and body regions to have different flow resistance.

3. Write down the modified Bernoulli equation and explain how it is used.

4. Define the *continuity principle* and explain one way it can be used.

5. Name the two main effects of a stenosis in an artery or heart valve that can be detected using Doppler.

6. Describe how arterial flow waveforms differ between high resistance and low resistance circulations.

7. Define the term *Resistance Index*. Describe how it is calculated and explain why it is independent of the Doppler angle.

8. Explain with the aid of diagrams what is meant by laminar flow and a parabolic velocity profile.

9. Define what is meant by the term *boundary layer separation*.

Chapter 10: Imaging performance and limitations

1. Define the term *spatial resolution*. Name the two principal technical factors that affect the spatial resolution.

2. Define *slice thickness* and explain why it is relatively poor when non-matrix probes are used.

3. Define the term *contrast resolution*. Name some of the technical factors that affect the contrast resolution.

4. Define the meaning of the term *temporal resolution*. Name two machine parameters that affect the temporal resolution.

5. Briefly discuss the effect of the depth of penetration, write zoom, image width, multiple focus, compound imaging, harmonic imaging and colour Doppler on the frame rate.

Chapter 11: Bioeffects and safety

1. Define the meaning of the terms *bioeffect* and *biohazard*.

2. Name the two major categories of biohazard in diagnostic ultrasound and an example of each.

3. Define the meaning of the terms *Thermal Index* and *Mechanical Index* and discuss the reason that they are displayed on the screen. Discuss the safe limits that have been agreed by international consensus.

4. Explain why there are three different Thermal Index values that can be displayed - TIs, TIb, TIc.

5. Discuss the difference between the temporal peak, pulse average and temporal average intensity values.

6. Define what is meant by the term *duty factor* (or duty cycle). Indicate how it relates to the difference between the pulse average and temporal average intensities.

7. Summarise the main points of the Australasian Society for Ultrasound in Medicine's safety policies (or select the policies of another national or international organisation). Describe how these should be put into practice.

Chapter 12: Additional modes and capabilities

1. Describe how compound imaging differs from standard imaging. Explain the purpose of using compound imaging, and name its primary advantages and limitations.

2. Explain why the echoes from soft tissues can be distorted and therefore contain energy at harmonic frequencies.

3. Explain the principle on which tissue harmonic imaging is based. List four ways in which image quality is improved.

4. Explain why echoes from contrast bubbles are distorted and therefore contain energy at harmonic frequencies. Describe why harmonic imaging is used with contrast agents.

5. Name two limitations of harmonic imaging when compared with standard imaging.

6. Explain the nature of ultrasound contrast agents and briefly describe three clinical applications of them.

7. Describe, with the aid of diagrams, how three-dimensional ultrasound imaging is achieved. Name two different techniques used to sweep the imaging scan plane through a volume of tissue.

8. Name three different methods used to display 3D ultrasound information.

9. Define the term *speckle tracking* and explain how the machine achieves it.

10. Explain what is meant by *strain* and *strain imaging*. How does the ultrasound machine measure strain?

11. What is meant by ultrasound *elastography*? Name two different methods by which ultrasound machines do elastography.

12. Briefly explain what is meant by *synthetic aperture imaging*. Name two advantages that it has over conventional imaging.

13. What is meant by the term *Artificial Intelligence (AI)*? Name two ways ultrasound machines may use AI.

14. Explain with the aid of diagrams how extended field of view imaging is achieved. Name one practical issue that the user needs to be aware of when doing an extended field of view scan.

18

C: Model answers

The following answers list the main points that should be mentioned in your answer. They are not exhaustive and the order in which points are listed is not necessarily significant. Where the question asks for one or more diagrams, refer to the textbook *The Physics and Technology of Diagnostic Ultrasound: A Practitioner's Guide (Second Edition)* by Robert Gill for relevant diagrams.

Chapter 2: Ultrasound interaction with tissue

1. Energy = work (measured in Joules).
   Power = energy per second (measured in Watts).
   Intensity = power per unit area (measured in Watts/cm²).
   Intensity is the most relevant parameter when you are assessing patient exposure since it takes into account whether the ultrasound energy is spread out through a significant volume of tissue or whether it is concentrated in a small volume (e.g. due to focussing or when the beam is stationary as in pulsed Doppler).

2. Wavelength = physical distance occupied by one cycle.
   Period = time taken for one cycle.
   Pulse duration = time taken for one transmit pulse.

3. If f = 4.5 MHz then
$$\lambda = \frac{c}{f}$$
$$= \frac{1.54 \times 10^3 \text{ m/s}}{4.5 \times 10^6 \text{ Hz}}$$
$$= 0.34 \times 10^{-3} \text{ m}$$
$$= 0.34 \text{ mm}$$

If f = 12.5 MHz then

$$\lambda = \frac{1.54 \times 10^3 \text{ m/s}}{12.5 \times 10^6 \text{ Hz}}$$
$$= 0.12 \times 10^{-3} \text{ m}$$
$$= 0.12 \text{ mm}$$

4. Amplitude = amount by which the pressure in the patient's tissues is increased and decreased (relative to normal pressure) due to the presence of the ultrasound wave.
    Intensity is proportional to (amplitude^2).

5. Absorption of ultrasound energy by the tissues.
    Reflection of ultrasound energy.
    Scattering of ultrasound energy.
    Defocussing of the ultrasound beam, as occurs in edge shadowing.
    Absorption is the dominant factor in normal tissue.

6. Attenuation:
    $$attenuation = \alpha \times L \times f$$
    $$= 0.5 \times 20 \times 3.5$$
    $$= 35 \text{ dB}$$

    (note: this means the intensity is reduced by a factor of $10^{3.5} = 3,200$)

7.
    $$attenuation = 10 \times \log\left(29,700\right) = 45 \text{ dB}$$

8. Depth of penetration = maximum depth from which echoes can be detected and displayed. Echoes from greater depths are too weak to display. We can show that
    $$penetration \times frequency = P \times f = constant$$
    So depth of penetration is inversely proportional to frequency.

9. Using the above equation
    $$P_1 \times f_1 = P_2 \times f_2$$
    therefore:
    $$8 \text{ cm} \times 10 \text{ MHz} = P_2 \times 4 \text{ MHz}$$
    and so:
    $$P_2 = \frac{8 \times 10}{4} = 20 \text{ cm}$$

10. Reflection is the term used when some or all of the energy incident on a relatively smooth surface does not pass through the surface but instead travels back into the first tissue again. The direction in which it travels is easily determined, since the incident angle is equal to the reflection angle. (Note these angles are measured between the direction of travel of the ultrasound and a line perpendicular to the interface.)
    When the incident angle is 90° the reflected ultrasound (i.e. the echo from the interface) travels directly back to the probe and so it will be detected and displayed. When the incident angle is not 90° the reflected ultrasound travels away from the probe and it will not be detected. The interface will then be seen in the image only if it also scatters sufficient ultrasound to be visible.

11. The fraction of the ultrasound power that is reflected is defined as the reflection coefficient (R). Since the factor of 10^6 is in the top and bottom of the equation it will cancel out, so R can be calculated as follows:

$$R = \frac{(z_2 - z_1)^2}{(z_2 + z_1)^2}$$
$$= \frac{(1.50 - 1.55)^2}{(1.50 + 1.55)^2}$$
$$= 0.00027$$
$$= 0.027\%$$

12. Energy cannot be created or destroyed, so the sum of the reflected and transmitted power must be equal to the power incident on the interface. It follows that the sum of the reflection and transmission coefficients must be 1. Therefore the fraction of the power that is transmitted through the interface is:

$$T = 1 - 0.015 = 0.985 = 98.5\%$$

13. Reflection is described in the answer to question 10 above. Scattering is the term used when ultrasound interacts with a small object. A fraction of the ultrasound energy is 'scattered', with the remaining energy continuing on unchanged. The scattered energy is distributed in all directions. This is the major difference between reflection and scattering.
    Reflection gives rise to linear echoes that represent the reflecting tissue interface. Often these echoes are quite strong,

especially when the ultrasound is at perpendicular incidence to the tissue interface.

The echoes produced by scattering are generally lower in level. In addition, the interaction between multiple scattered echoes produces the characteristic 'speckle' appearance seen in ultrasound images of soft tissue.

14. See the answer to question 13 above.
    Frame averaging (persistence) and compound imaging both tend to reduce speckle.

15. Refraction is the term used when the direction of the ultrasound beam changes as a result of passing through an interface between two tissues that have different ultrasound propagation speeds. The change of angle is calculated using Snell's Law:

$$\frac{\sin \theta_i}{c_1} = \frac{\sin \theta_t}{c_2}$$

where θ_i and θ_t are the incident angle and transmitted angle respectively and c_1 and c_2 are the propagation speeds in the first and second tissues respectively.

16. Using Snell's Law (see the answer to question 15 above):

$$\frac{\sin 35°}{1470} = \frac{\sin \theta_t}{1550}$$

$$\sin \theta_t = \frac{1550}{1470} \times \sin 35°$$

$$= 0.605$$

and so

$$\theta_t = \sin^{-1} 0.605 = 37.2°$$

Yes, this interface does have a critical angle (defined as the incident angle for which $\theta_t = 90°$) since $c_2 > c_1$.

17. As mentioned above, the critical angle is the value of the incident angle for which the transmitted angle is 90°. This can only occur when the propagation speed in the second tissue is greater than the propagation speed in the first tissue.

    When the incident angle exceeds the critical angle, the ultrasound is reflected by the interface and does not pass through it. This causes artifacts such as edge shadowing and mirror image, and it prevents the second tissue from being imaged.

Chapter 3: Pulsed ultrasound and imaging

1. (a) Assume the echo comes from the centre-line of the ultrasound beam.
    (b) Depth = (c × t)/2 where c is the propagation speed and t is the echo arrival time (i.e. the difference between when the transmit pulse was transmitted and when the echo was received).

2. Using the equation in the answer to question 1:
$$depth = \frac{1.54 \times 10^3 \times 230 \times 10^{-6}}{2}$$
$$= 1.77 \times 10^{-1} \text{ m}$$
$$= 17.7 \text{ cm}$$

3. Pulse duration.
    Two objects will be 'resolved' (i.e. seen in the image as two separate objects) if their echoes do not overlap in time. This will be the case as long as the arrival times of the two echoes differ by at least an amount equal to the pulse duration (τ). Using the relationship between echo arrival time and depth cited in the answer to question 1 above, this means that the difference in depth between the two objects must be greater than (c × τ)/2 which is therefore the formula for the axial resolution.

4. PRF = number of transmit pulses per second.
    Machine must wait until all detectable echoes have been received before it can transmit again. The deeper it is penetrating the longer it must wait and so the lower the PRF will be.
    The last detectable echo comes from the penetration depth P. It will be received at a time 2P/c after the last transmit pulse. If the machine transmits again at this time (since this will maximise the PRF) the period of the PRF will be 2P/c and so the PRF will be c/2P. Thus the relationship between penetration and PRF is a reciprocal one.

5. PRF:
$$PRF = \frac{c}{2P}$$
$$= \frac{1.54 \times 10^5 \text{ cm/s}}{2 \times 18 \text{ cm}}$$
$$= 4300 \text{ Hz} = 4.3 \text{ kHz}$$

and

$$PRF = \frac{1.54 \times 10^5 \text{ cm/s}}{2 \times 4 \text{ cm}}$$
$$= 19300 \text{ Hz} = 19.3 \text{ kHz}$$

6. Continuing from the answer to question 4 above, let us assume it requires N transmit pulses to create each image. Then the number of images per second is simply the PRF divided by N. Using the result above:

$$FR = \frac{PRF}{N} = \frac{c}{2PN}$$

This is often written as:

$$FR \times P \times N = \frac{c}{2}$$

7. Frame rate:

$$FR = \frac{(1.54 \times 10^5 \text{ cm/s})}{(2 \times 10 \text{ cm} \times 200)} = 38.5 \text{ Hz}$$

8. With both the linear and curved probes, a group of transducer elements form the aperture that is used to create the beam. The beam is moved along the probe by incrementing the aperture, generally by dropping off one element and replacing it with another at the other end of the aperture. For example, elements 1 - 128 are used for the first beam, 2 - 129 for the second etc. The beam is normally perpendicular to the transducer face and so a linear array scans a rectangular region and a curved array scans a region that spreads out with depth.

   The phased array uses all the transducer elements at once. It keeps the starting point of the beam fixed in the centre of the transducer and steers the beam in different directions to produce a sector scan.

9. See description in the answer to question 8.

10. Linear array: peripheral vascular, neck, breast.
    Curved array: abdomen.
    Phased array: heart.

11. B-mode: standard two-dimensional image where the brightness at a point indicates the echo strength.
    M-mode: the beam is kept in a fixed position and it produces an image which is a single line of echoes along this beam; this

image is swept slowly across the screen to create a time-motion display of tissue movement.

A-mode: Again the beam is kept in a fixed position; in this mode the amplitude is displayed as vertical deflection on a graph, with time (i.e. depth) on the horizontal axis.

Chapter 4: Transducers and focussing

1. The term piezoelectric is used to describe materials that produce an electrical voltage when they are deformed (e.g. squeezed or expanded). The inverse piezoelectric effect is where the dimensions of the material change in response to an applied electrical voltage.

2. The transducer is made of piezoelectric material (see the answer to question 1 above). An electrical voltage that oscillates at the ultrasound frequency is applied between the top and bottom faces of the transducer. The transducer oscillates in thickness as a result, creating pressure variations in the tissues that become an ultrasound wave.

3. When an echo reaches the transducer it produces oscillations in pressure that cause the transducer thickness to oscillate. This causes an oscillating electrical voltage to be produced between the front and back faces of the transducer. This is the echo signal.

4. Transducer: converts electrical oscillations to ultrasound waves and vice versa.
    Backing layer: absorbs ultrasound energy that would otherwise be transmitted into the probe case, causing misleading echoes.
    Matching layer: a thin layer between the transducer and tissues that improves the flow of ultrasound energy between them since it makes it appear that they have the same acoustic impedance.
    Lens: a curved acoustic lens runs the length of the probe to focus the slice thickness dimension of the ultrasound beam; this is not required if a matrix probe is used.
    Gel: used as the coupling medium between the probe and the patient, its purpose is to eliminate air from the space between them to improve the transmission of ultrasound.
    Electronics: most probes contain electronics for selecting and switching between the transducer elements.

5. When the point reflector or scatterer is inside the beam it will create an echo and it will be displayed in the image as if it was on the centre-line of the beam. Since the beam steps by small increments (far less than the beam width in size), the point will

remain within the beam for several successive beam positions. It will be displayed each time in a different lateral position as the beam moves. Thus the point will be imaged multiple times at the correct depth but over a range of lateral positions. The images will join together to form a line in the image, equal in width to the beam width at that depth.

6. Spatial resolution is a measure of the sharpness of the image. It is defined as the minimum separation required between two point objects for them to be 'resolved' (i.e. seen as two separate objects) in the image.
Beam width (determines lateral resolution).
Pulse duration (determines axial resolution).

7. The diffraction limit defines the width that the beam will have when it is focussed. The limit consists of a pair of diverging lines. The beam is wider than this limit except at the focal depth, where it touches the limit.
Aperture.
Frequency (or wavelength).

8. The divergence angle is:

$$\theta = sin^{-1}(\frac{1.22 \times \lambda}{A})$$

The wavelength is

$$\lambda = \frac{c}{f}$$
$$= \frac{1.54 \times 10^5 \text{ cm/s}}{4.5 \times 10^6 \text{ Hz}}$$
$$= 0.034 \text{ cm}$$

and so

$$\theta = sin^{-1}(\frac{1.22 \times 0.034 \text{ cm}}{2.5 \text{ cm}})$$
$$= 0.95°$$

9. The beam width at focus is:

$$bw_f = \frac{(2.44 \times \lambda \times F)}{A}$$

where F is the focal depth. The wavelength is:

$$\lambda = \frac{(1.54 \times 10^5 \text{ cm/s})}{(5 \times 10^6 \text{ Hz})} = 0.031 \text{ cm}$$

and so:

$$bw_f = \frac{(2.44 \times 0.031 \text{ cm} \times 4 \text{ cm})}{1 \text{ cm}}$$

$$= 0.30 \text{ cm}$$

$$= 3.0 \text{ mm}$$

10. These probes are not cut up into elements across the width. Instead, each transducer element runs across the full width of the transducer. Thus the machine cannot use electronic focussing. Instead it simply uses a fixed lens to focus in this direction. The focus is inferior to the focus within the scan plane because (a) the aperture is smaller in this direction, and (b) the depth of focus is fixed, not adjustable.

11. A matrix array transducer is a transducer that has been sliced both lengthwise and across the width, creating a very large number (generally thousands) of small transducer elements. Each of the elements can transmit and receive ultrasound independent of the others.

    Matrix arrays can focus electronically across the width of the transducer (the slice thickness direction), reducing the slice thickness.

    They can also steer the beam in the slice thickness (elevation plane) direction. This sweeps the scan plane through a volume of tissue, as required for 3D imaging.

12. A group of transducer elements is used to transmit the ultrasound pulse. The elements operate individually as transmitters. The machine uses electronic delays to modify the exact time that each element transmits. This is done in such way that the transmitted pulses from all of the elements arrive at the point of focus at the same instant. This guarantees that the intensity is maximised at that point and therefore the beam width is minimised.

    Similarly when the transducer elements are receiving ultrasound echoes, the echo signal from each element is delayed by a different amount before the signals are added together to produce the echo signal that the machine will process and display. The purpose of this is to ensure that echoes from the point of focus will be aligned in time so they will add to give the maximum possible intensity.

13. When the ultrasound machine is receiving echoes it continually adjusts the depth of focus to match the depth from which the echoes are returning. For example, 13 μs after the transmit pulse, the echoes will be coming from a depth of 1 cm and so the machine adjusts the receive beam former delays to focus at a depth of 1 cm. Thus the depth of focus is swept deeper and deeper with time to track the depth that the echoes are coming from.
This is not possible when the machine transmits. Once the ultrasound has been transmitted into the patient, it is no longer under the machine's control and so it will focus at one specific depth determined by the transmit beam former delays.

14. The machine transmits with focus depth 1 and receives echoes from the tissues around this depth. Without moving the beam it then transmits again with focus depth 2 and receives echoes around this depth. This continues for each of the focus depths that are in use. The beam is then moved to its next position and the process starts again. In this way the image is acquired as a series of strips at different depths (note these strips must overlap). The strips are then merged to form a single complete image which is then saved in the image memory and displayed.
Advantage: better and more uniform focus over the entire depth of the image.
Disadvantage: the frame rate will be reduced in proportion to the number of focusses used.

15. Pulsed Doppler.
Colour Doppler.
Compound imaging.
Trapezoidal scanning.

Chapter 5: Ultrasound instrumentation

1. Transmitter: frequency, output power, Doppler velocity scale (varies the PRF).
   Beam-formers: focus depth, single/multiple focus, field of view, Doppler beam steering.
   Amplifier: gain.

2. The term TGC describes the way the machine increases the gain as the echoes come from increasing depth.
   This is required to compensate for the attenuation of ultrasound in tissue which would otherwise cause the echoes to grow weaker with depth in the image.

3. The term dynamic range is used to describe the overall range of echo intensities used to make the image. It is measured by calculating the ratio of the strongest echo intensity to the weakest, usually expressed in decibels.
   If changes in echo intensity were translated directly into changes in display intensity (e.g. if a doubling of echo intensity meant the brightness of the echo on the display doubled) the dynamic range that could be displayed would be limited to just 25 - 30 dB by the display. Typically the echoes have a dynamic range of 60 dB or more, so much information would be lost.
   Instead the machine has a dynamic range compression function which maps variations in echo intensity into variations in display intensity. Generally, this is designed to show variations in the strength of soft tissue echoes at the expense of making all the higher-level echoes from tissue interfaces look much the same.
   (a) Large dynamic range: low-contrast or 'soft' looking image, soft tissue has a fairly uniform appearance, artifact echoes may be quite evident.
   (b) Small dynamic range: more 'contrasty' coarse-looking image, lower level echoes lost, may be areas of dropout, reduced penetration, less artifacts visible.

4. The scan converter takes the echoes as they come from the machine's processing and determines where to display them in the image memory. Thus it 'converts' between the scan format of the probe and the rectangular grid used in the image memory. It also fills in any empty pixels which do not have echo

data written to them with values calculated using mathematical interpolation.

5. Pre-processing refers to functions that alter the echo data before it is stored in the image memory. While this could include all the prior processing of the echoes (gain, TGC, dynamic range etc) it is usually confined to processes that occur immediately before the echoes are stored. Since pre-processing functions alter the information that is stored in the image memory, their effect is irreversible. Examples include:
Frame averaging (or persistence) - the current image is combined with the previous image before being stored in the image memory. This reduces speckle.
Compound imaging - several images are created with the beam steered in different directions and these are combined before the final composite image is stored in the image memory. This reduces speckle and improves tissue boundaries.
Extended field of view scanning - as the probe is moved, any new image information is merged with the existing image and so an image covering a larger field of view is built up over time. Other examples include depth and write zoom.

6. Post-processing refers to functions that alter the image data as it is being read out from the image memory for display. The data in the image memory is not altered, and so these functions can be turned on and off and their effects are reversible. Examples include:
Read zoom - the stored image is magnified so only a selected portion is displayed on the screen.
Colour mapping - the grey scale image is converted into a coloured image.
Post-processing curves - these manipulate the way echo strength is mapped to brightness on the display.
Measurements can also be thought of as post-processing.

7. When the freeze button is activated, the cine-loop function allows the user to review the images acquired over a number of seconds immediately prior to the freeze button being hit. This allows the user to activate freeze after seeing an image of interest, then scroll back to find it.

Chapter 6: Image artifacts

1. An artifact is any appearance in the image that does not accurately reflect the tissues being imaged. A tissue may be wrongly displayed (e.g. distorted, displaced, multiplied), missing from the image or there may be echoes in the image that do not represent real tissues.

2. Shadowing is caused by the presence of tissues with higher attenuation than the surrounding tissues. The transmitted ultrasound that passes through these tissues is reduced in amplitude by more than would otherwise be expected. Similarly the returning echoes that pass through these tissues are attenuated more than would otherwise be expected. The result is that the echoes coming from deep to the strongly attenuating region are lower in level than they would be if the attenuation in this region was normal, and so the image is darker in this region than elsewhere.
    Gallstones often cause shadowing.
    Shadowing is often useful, since it draws the user's attention to areas that have abnormally high attenuation such as stones. Where shadowing is a problem, alternative acoustic windows need to be found to image the region of interest.

3. Enhancement is caused by the presence of tissues with lower attenuation than the surrounding tissues. The transmitted ultrasound that passes through these tissues is higher in amplitude than would otherwise be expected. Similarly the returning echoes that pass through these tissues are attenuated less than would otherwise be expected. The result is that the echoes coming from deep to the low attenuating region are higher in level than they would be if the attenuation in this region was normal, and so the image is brighter in this region than elsewhere.
    Liquid filled areas such as cysts often cause enhancement.
    Enhancement is often useful, since it draws the user's attention to areas that have abnormally low attenuation.
    Enhancement is not generally seen as a problem. However it may at times cause excessive brightness in the deeper tissues in which case the TGC or gain should be adjusted to optimise the image in this area.

4. When the ultrasound beam strikes the edge of a curved structure it is reflected and/or refracted, causing (a) deflection of the beam and (b) defocussing, since the beam spreads out more rapidly than it would otherwise have done. The result of the defocussing is that the intensity is reduced, causing the echoes to be reduced in intensity. These echoes are therefore darker in the image than they would otherwise have been, causing a shadow. Since the machine does not know that the beam has been deflected, the shadow is displayed in line with the direction that the machine transmitted the beam.
    Edge shadowing can be caused by the edges of blood vessels when they are viewed in cross-section.
    Edge shadowing is not generally thought to be useful or a problem.

5. If ultrasound travels through a region of tissue with propagation speed significantly different to the value that the machine assumes (1540 m/s), the depth of objects deep to this region will be incorrectly displayed. If the region has a lower propagation speed, the tissues will appear to be deeper than their true depth. If it has a higher propagation speed, the tissues will appear to be less deep than their true position.
    This artifact can occur if there is a localised area of fat in the liver.
    It may occasionally be useful in providing additional information. Usually the distortion of the image is small, but occasionally this artifact can cause significant measurement errors.

6. Ultrasound can reflect back and forth between reflective parallel objects, giving rise to multiple equally spaced echoes in the image. This reverberation of the ultrasound may be between a tissue interface and the transducer face (especially if there is not enough coupling gel) or between two tissue interfaces. Note that both the tissue interfaces and the soft tissues between them will be replicated multiple times at increasing depth in the image.
    Reverberation is common in the abdominal wall.
    It is not usually regarded as useful. It can be reduced by angling the probe so that the ultrasound beam is not at perpendicular incidence to the tissues in question.

7. When ultrasound strikes small gas bubbles it reverberates within and between the bubbles, giving rise to a series of relatively strong echoes (since bubbles are very effective scatterers of ultrasound, especially when they resonate). Since bubbles are good scatterers of ultrasound, little energy is lost in the scattering process. Also, bubbles are generally suspended in liquid, which has low attenuation. So ringdown artifact causes a bright band of echoes in the image which often extend to the full depth of the image. A needle (or other metallic structure) can produce a similar appearance as the ultrasound reverberates.
Ringdown artifact is often useful since it can highlight the presence of even small amounts of gas. It is also at times a major problem, since it prevents imaging of the tissues deep to the gas. An alternative window must then be found, and in some cases invasive probes must be used.

8. When ultrasound strikes small calcifications or other crystalline materials, it reverberates within them, causing a series of relatively bright echoes. A significant amount of energy is lost, so the bright echoes only last for a relatively short distance before they fade away. The result is a 'tail' extending in depth from the initial echo for a limited distance.
Microcalcifications can cause comet-tail artifact.
This artifact is often useful, since it provides extra information about the structures causing it. It is unlikely to be a problem.

9. Ultrasound machines are designed to make sure they do not transmit until all detectable echoes from the previous transmit pulse have been received. Occasionally they fail to achieve this. This may happen if the ultrasound is travelling through a significant volume of fluid (such as amniotic fluid or urine in a full bladder), since the low attenuation of fluid means that the penetration depth is larger than usual.
When this happens, echoes generated by the previous transmit pulse are received after the next pulse has been transmitted. The machine must assume these echoes were generated by the new pulse, so they are displayed at a much more superficial depth than they should be.
This is not a common artifact and it is not generally a problem when it does occur, since it is often easy to recognise as an artifact.

10. As the ultrasound beam scans, it displays each point in the image as a linear structure (as described in Chapter 4, question 5). This effect is easily seen when a tissue-liquid interface (such as the gallbladder wall) is viewed at an angle other than perpendicular incidence.

    Beam width artifact is not useful. Instead, it degrades image resolution and reduces the accuracy of measurements.

    It can be reduced by focussing the beam better at the point of interest (by placing the focus marker at that depth and using a higher frequency) and by ensuring the beam is perpendicular to tissue interfaces as much as possible.

11. When the ultrasound beam is generated by the machine there will always be lower-level beams either side of the main beam. These are referred to as sidelobes. As the main beam scans through the tissues creating the ultrasound image, the sidelobes are also scanning the tissues. Even though the transmit intensity and receive efficiency of the sidelobes is much lower than the main beam, strong reflectors can sometimes produce detectable echoes.

    These will be incorrectly positioned in the image since they will be displayed as though they have come from the main beam. A typical situation where sidelobe artifacts can occur is where the fetal head is scanned and sidelobe artifacts are visible either side of the head in the echo-free amniotic fluid.

    This is not a useful artifact, and neither is it generally a problem unless the user fails to recognise it as a sidelobe artifact. Often there is little the user can do to reduce or eliminate this artifact.

12. Usually we think of the ultrasound image as representing a thin slice through the patient's tissues. In fact the beam has significant 'thickness' in the direction across the width of the transducer and so the image is derived from a volume of tissue. This 'slice thickness' leads to tissues appearing in the image even when they are not in the central plane of the scanned volume. A typical situation where this occurs is when relatively small blood vessels (e.g. leg veins) are scanned in long axis. Echoes from the tissues either side of the vessel can be displayed as if they have come from within the vessel.

    This is not a useful artifact and it can be a problem. Selecting a suitable probe is important, since this can minimise the slice thickness. In the situation above, rotating the probe to image the vessel in short axis will cause the artifact to disappear.

13. Soft tissue contains a very large number of small structures that scatter ultrasound (for example, small blood vessels and lymphatic vessels). Each of these produces its own echo. When the echo signals are detected by the transducer, they add together to produce the echo signal that the machine displays. The amplitude of the echo signal depends on how the individuals add up. Sometimes they will add 'constructively' to produce a visible echo, at other times they will cancel each other out to produce an echo-free point in the image. The result is a random pattern of grey scale values known as 'speckle'.

14. Refraction (alteration in the direction of the ultrasound beam) occurs whenever the beam passes through an interface between two tissues with different propagation speeds at an angle other than perpendicular incidence. The machine displays all echoes from deeper tissues as though the beam had not changed direction, and so the echoes will be misplaced in the image. The best-known example is when structures such as the superior mesenteric artery are doubled in the image when scanning transverse on the abdomen, due to refraction of the beam by the anterior abdominal wall muscles.
This is not a useful artifact. It may cause confusion in the specific situation described above. Moving the probe to the left or right of the centre-line of the body will generally eliminate the doubling effect.

15. Some or all of the ultrasound energy will be reflected when it strikes a tissue interface with a large impedance difference (e.g. soft tissue - air) or at an angle greater than the critical angle. The reflected beam will then continue to generate echoes and these can follow the same pathway as the transmitted pulse back to the probe. These echoes will therefore be displayed in the image, but they will be positioned incorrectly by the machine because it cannot take into account the reflection. Thus the echoes will be displayed deep to the reflecting interface, just like a mirror reflecting light. A typical example is when scanning the liver; the diaphragm normally represents an interface between soft tissue (liver) and air (air-filled lungs) and

so it is highly reflective. Liver tissue is therefore often mirrored and displayed deep to the diaphragm.

This is not a useful artifact. However, it can be important to recognise that it is occurring. Moving the probe will help to identify it as artifact.

16. Artifact echoes can be thought of being overlaid on top of the genuine ultrasound image. Even when they are not evident to the user, they reduce the user's ability to see small variations in soft tissue echo strength which might be indicative of pathology. In other words, artifact echoes reduce the contrast resolution.

Chapter 7: Doppler ultrasound

1. Ultrasound echoes have a slightly different frequency to the frequency that was transmitted if the tissues causing the echoes are moving relative to the transducer. This frequency difference is called the Doppler shift. The machine detects and displays the Doppler shift in various ways: as audible sound; as a spectral display; as a colour overlay on the grey scale image. Doppler ultrasound therefore provides information regarding the location and distribution of flowing blood and the speed at which it is moving and how this changes with time. It can also be used to assess tissue motion, for example in the heart.

2. The Doppler shift is proportional to the speed of movement of the blood relative to the probe. This is $v \cos \theta$ where v is the blood velocity and θ, referred to as the Doppler angle, is the angle between the ultrasound beam and the direction of blood flow. The Doppler shift is therefore proportional to $v \cos \theta$.

3. Often it is important to determine the blood velocity from the measured Doppler shift. This means that the value of the Doppler angle (θ) must be known. The user places the Doppler angle cursor parallel to the direction it is assumed the blood is flowing (generally parallel to the vessel walls in vascular applications). The machine then calculates the angle between this line and the beam direction. It is likely that the angle measured in this way may be incorrect by as much as a few degrees. When the Doppler angle is small, an error of a few degrees in determining the angle makes little difference to the calculation of the blood velocity from the Doppler shift. However, when the Doppler angle is large, a few degrees error in the angle will cause a substantial error in the velocity calculation. The general consensus is that Doppler angles greater than 60° will be associated with unacceptable errors and should not be used when velocity values will be measured.

4. Scattering from blood is weak compared to soft tissue, as shown by the lack of visible echoes from blood in grey scale images. To compensate for this, the machine increases the transmitted power in Doppler mode (generally by lengthening the transmit pulse duration). Furthermore, in pulsed Doppler the ultrasound

beam does not scan through the tissues but it remains stationary. The ultrasound energy is therefore concentrated in a small volume of tissue.

5. One transducer transmits continuous ultrasound and the second transducer continuously receives the echoes returning from the patient's tissues. The machine detects the difference between the transmitted and received frequencies (i.e. the Doppler shift). It uses quadrature detection to resolve the direction of flow relative to the probe (i.e. to determine whether the blood is flowing towards or away from the probe).
The sample volume is the region of overlap of the transmit and receive beams, and it may be quite large. Where an array transducer is used (e.g. in echocardiography) one group of transducer elements is used as the transmitter and a second group of elements as the receiver.

6. Horizontal axis: time (generally several seconds).
Vertical axis: velocity (calculated by the machine from the Doppler shift, measured in cm/s or m/s).
Positive velocity (conventionally above the zero line): flow towards the probe.
Negative velocity: flow away from the probe.

7. Nyquist Limit: the maximum Doppler shift that can be detected unambiguously, equal to one half of the Doppler PRF. Measuring unusually high velocity values can be very important in the diagnosis and grading of stenotic disease. Unfortunately, the Doppler shift often exceeds the Nyquist Limit in these cases, causing frequency aliasing. If the aliasing is not too severe, shifting the baseline of the display up or down can conceal the effect of the aliasing in the display, since this increases the display range in one direction.

8. The user places the sample volume (the region from which the Doppler signal is to be acquired) in the area of interest and adjusts the Doppler angle cursor to inform the machine of the flow direction. The machine selects an appropriate beam and calculates the depth of the sample volume and the Doppler angle. With the beam stationary (not scanning), the machine transmits pulses repeatedly down this beam and receives the echoes. It then 'range gates' the echoes at the depth of interest,

ignoring echoes from nearer and deeper tissues. It repeats this process around 128 times then combines the range gated echoes from these pulses. The result is the Doppler signal. It is wall-filtered to remove low Doppler shift signals likely to be caused by tissue movement. The Doppler signal is then spectrum-analysed and displayed. Since the Doppler angle is known, the machine calculates the blood velocity from the measured Doppler shift using the Doppler equation.

9.   [See textbook Figure 7.15.]

10. The upper and lower limits of the spectral display are equal to the Nyquist Limit (half the Doppler PRF). Thus the overall range of velocity values that can be displayed ranges from -PRF/2 to +PRF/2, a total span equal to the PRF. Even when the baseline of the display is shifted, the total display range is still equal to the PRF. Thus changing the PRF will alter the velocity scale.

11. The user positions the colour box, adjusting its size and the beam steering angle. The machine acquires a complete grey scale image (with the beams directed in the usual imaging direction) and then it acquires colour Doppler information for the region within the colour box. For each beam position the machine transmits and receives 8 or so times to acquire the colour Doppler information.
A large number of closely spaced range gates are positioned along the beam covering the tissues within the colour box. The range-gated echoes from each range gate for a given beam position are then analysed to produce the colour information for that point in the image. The calculation produces three numbers - mean Doppler shift, variance and Doppler power. The machine also determines whether the point is tissue or blood. If it is blood, the mean Doppler shift is converted to a colour and displayed. If it is tissue, the grey scale information is displayed instead.
The colour bar shows how the Doppler shift is converted to colours. Generally red (above the baseline) represents flow towards the probe and blue flow away from the probe. The shade of red or blue varies depending on the Doppler shift frequency.

The grey scale (or tissue) threshold value is usually displayed on a grey scale bar adjacent to the colour bar. A grey scale value above this level at any given point will override colour.

12. As discussed in the answer to question 11, the machine collects 8 or so echo samples for each range gate for each beam position. The echoes are processed using autocorrelation, a mathematical algorithm. The result is three numbers for each range gate in the colour box - mean Doppler shift, variance (a measure of the range of Doppler shifts) and Doppler power.
The machine uses these parameters, plus the grey scale value in the image at that point, to determine whether the point is more likely to be blood (in which case the colour is displayed and the grey scale suppressed) or tissue (in which case the grey scale is displayed and the colour is suppressed).

13. Because only eight or so pulses are used to acquire the colour Doppler information at each point, colour Doppler is inferior to pulsed Doppler in several ways.
Less accurate determination of velocity.
Less effective wall filtering.
Less able to detect slow-moving blood.

14. Mean Doppler shift - the average Doppler shift for blood in the sample volume. This is the parameter that is usually coded in colour in the image.
Variance - a measure of spectral broadening. Used in the tissue-blood discrimination process. In addition, areas of abnormally high variance can be highlighted in the image using a secondary colour (commonly green) to identify regions that are likely to contain disturbed or turbulent flow.
Doppler power - a measure of the strength of the Doppler signal. Also used in the tissue-blood discrimination process. Furthermore, in power-mode colour Doppler this is the parameter that is coded in colour, rather than the Doppler shift.

15. Variance - this is a measure of the range of Doppler shifts at a given point (similar to the the width of the Doppler spectrum in spectral Doppler). If the variance is abnormally large this indicates that there is increased spectral broadening at that point which strongly suggests the presence of disturbed or turbulent flow.

16. Generally the Doppler angle will be different at different points within the colour box. This means that differences in the colour displayed at different points may be due to differences in velocity or angle or both. Thus it can be very difficult to determine from the colour Doppler image why different colour values are displayed at different points in the image.

17. In power mode colour Doppler, the Doppler shift is not displayed. Instead the power (strength) of the Doppler signal at each point determines what colour will be displayed. In some machines the colour can also be coded to indicate whether the blood is flowing towards or away from the probe.
Power mode colour Doppler can be useful for assessing vascularity in a region and searching for trickle flow in a vessel (e.g. in a severely stenosed artery that may or may not be totally occluded).
Advantages - more sensitive than standard colour Doppler (i.e. able to detect weaker Doppler signals), better able to detect slow flow, absence of confusing Doppler shift information when assessing vascularity, not affected by aliasing, able to detect flow even when the Doppler angle is 90°.
Disadvantages - no velocity information, information about flow direction relative to probe not always available.

Chapter 8: Doppler artifacts

1. The maximum Doppler shift that can be detected unambiguously is equal to one half of the Doppler PRF. This is known as the Nyquist Limit. When the Doppler shift exceeds the Nyquist Limit, frequency aliasing occurs (diagram).
To reduce or eliminate aliasing:
Increase the velocity scale (this increases the PRF).
Shift the baseline.
Reduce the ultrasound frequency.
Find a window which reduces the depth of the sample volume so the PRF can be increased further.
Increase the Doppler angle.
Use continuous wave Doppler.

2. For both imaging and Doppler, the ultrasound machine must wait until all detectable echoes have returned to the probe before it transmits the next transmit pulse. This means that the Pulse Repetition Frequency (PRF) is limited by the machine's depth of penetration. The deeper the penetration the lower the PRF must be.
In Pulsed Doppler it is sometimes necessary to increase the PRF beyond this limit to eliminate Doppler frequency aliasing. This is called 'high PRF Doppler'. In this case, a form of range ambiguity occurs, causing a second (unwanted) sample volume to appear at a different depth.

3. Intrinsic spectral broadening refers to a spread of Doppler shift values above and below the 'true' Doppler shift. This is caused by the fact that the ultrasound is transmitted and received using an area on the face of the transducer (the 'aperture'). As a consequence, a range of different paths are followed by the ultrasound on both transmission and reception. Each of these paths has a different angle relative to the direction of blood flow and so there is a range of Doppler angles rather than a single angle. As a result, an increased range of Doppler shifts is detected and displayed.
A consequence of this spectral broadening is that when velocity values are measured from the envelope of the spectral display (the usual way to make velocity measurements) the velocity will be over-estimated.

4. In some situations the spectral display is identical above and below the zero flow baseline. This is called a spectral mirror artifact. It generally occurs when the Doppler angle is 90°. It is in fact a form of intrinsic spectral broadening (see answer to question 3 above) and so it is caused by the finite aperture size. Occasionally a similar appearance can occur when the Doppler gain is too high, since signal overload can cause the machine's ability to determine whether flow is towards or away from the probe to break down.

5. Wall thump is a term used to describe interference caused by Doppler signals coming from moving blood vessel walls and cardiac structures.

6. Colour Doppler is based on the same range-gating principle as pulsed Doppler. This means that the Doppler shift information is 'sampled' once per transmit pulse, and thus at a rate equal to the Doppler PRF. If the Doppler shift exceeds the Nyquist Limit (equal to half the PRF) then an incorrect Doppler shift value will be detected and displayed.
    To reduce frequency aliasing in colour Doppler:
    Increase the colour velocity scale (this increases the PRF).
    Shift the colour baseline in the appropriate direction.
    Reduce the ultrasound frequency.

7. Sometimes objects such as gallstones can cause a rapidly changing Doppler shift value to be displayed in colour Doppler, causing a 'twinkling' appearance. Generally this means that the object has a crystalline structure.

Chapter 9: Haemodynamic concepts

1. The volume of blood that flows through a given region of the body is determined by the arterial-venous pressure difference (the driving force) and the flow resistance of the region. Flow resistance is a measure of how difficult it is to push blood through the region. Low resistance regions (such as the kidneys) require high blood flow rates while high resistance regions (such as the skin, or a limb at rest) require relatively low flow rates.

2. The flow resistance of most tissues is determined largely by the small arteries (arterioles, often called the 'resistance vessels'). If a large number of these are connected in parallel, the resistance is relatively low. Many tissues modulate their resistance depending on need (e.g. the arteries supplying the gut, or the leg muscles) by opening and closing some of these small vessels.

3. When blood flows through a narrowed region (a stenosis) there is a significant drop in blood pressure.
   The pressure drop across a stenosis (ΔP) can be estimated using the modified Bernoulli equation
   where v_2 is the blood velocity distal to the stenosis. This

   $$\Delta P = 4v_2^2$$

   equation assumes that v_1, the velocity proximal to the stenosis, is much smaller than v_2.

4. Blood is conserved (it is neither created not destroyed) as it flows through a given circulation. This means that the flow rates at two different points must be equal. This can be expressed as
   where v_1 and v_2 are the mean blood velocities at two different

   $$v_1 \times A_1 = v_2 \times A_2$$

   points and A_1 and A_2 are the corresponding lumen areas. Measuring the velocities at both points and measuring the lumen area at one point allows the area to be calculated for the second point

   $$A_2 = A_1 \times \frac{v_1}{v_2}$$

5. Increased blood velocity leading to an increased Doppler shift. Disturbed flow and turbulence immediately downstream from the stenosis, leading to increased spectral broadening.

6. High resistance arterial waveforms tend to have a relatively short sharp systolic peak, with flow then dropping rapidly (and often going negative for a short time). There may then be a short period of forward flow, after which flow will cease until the next heart contraction.
Low resistance arterial waveforms generally have rather slower systolic acceleration and continuous flow throughout the diastolic phase of the heart cycle.

7. The Resistance Index (RI) is calculated as (S-D)/S where S is the peak systolic flow velocity (or Doppler shift) and D is the minimum diastolic flow velocity (or Doppler shift). Both Doppler shift values are proportional to cos θ and so the cos θ term disappears when the ratio is taken. Therefore the ratio is independent of the Doppler angle.

8. (See the diagrams in the Velocity Profile section in the textbook). Laminar flow refers to the way the blood moves at different velocities at different points across the cross-section of the vessel. In general it moves slowly near the walls, with the velocity steadily increasing towards the centre of the vessel. The fastest flow is in the centre. This distribution of flow velocities is caused by frictional forces between the blood and the vessel wall, and between the blood at different points in the vessel.
In a straight vessel with steady flow, the blood velocity distribution across the diameter of the vessel (the 'velocity profile') is parabolic in form.

9. When rapidly flowing blood travels around a bend (as it does, for example, at the carotid bifurcation) the layer of blood adjacent to the vessel wall (the 'boundary layer') may fail to make the turn properly and so it loses contact with the wall. This is called boundary layer separation and it is due to the momentum of the blood (i.e. its tendency to continue in a straight line if possible).

Chapter 10: Imaging performance and limitations

1. Spatial resolution is a measure of the sharpness of the image, defined as the minimum separation required between two point objects for them to be 'resolved' (i.e. seen as two separate objects) in the image.
    Beam width (determines lateral resolution).
    Pulse duration (determines axial resolution).

2. Normally when we use the term 'beam width' we are referring to the beam width in the scan plane. However, the beam is three dimensional and it also has beam width in the direction perpendicular to the scan plane (the 'elevation plane' or 'slice thickness').
    For non-matrix probes, the focussing in this direction is less effective than focussing within the scan plane because the aperture size (in this case the width of the transducer) is smaller.

3. Contrast resolution refers to the user's ability to identify minor variations in echogenicity in soft tissue (for example, when there is a mass in an otherwise homogeneous organ such as the liver). Factors affecting the contrast resolution include:
    The presence of artifact echoes.
    The presence of speckle.
    Non-optimised gain, TGC and dynamic range settings.

4. Temporal resolution refers to the machine's ability to clearly image moving tissues.
    The frame rate (number of images per second) is the major factor affecting temporal resolution. It needs to be sufficiently high.
    Frame averaging (persistence) degrades temporal resolution, so it should be minimal when highly dynamic tissues (such as the heart) are scanned.

5. The frame rate depends on several factors: the depth of penetration P; the number of transmit pulses required to generate each line in the image E; the number of lines in each image L; the number of simultaneous beams the machine can generate M.

The maximum possible frame rate (FR) is related to these parameters as follows:

$$\frac{(FR \times P \times L \times E)}{M} = \frac{c}{2}$$

where c is the propagation speed. This relationship explains why:

Increasing the depth of penetration (P) reduces the frame rate. Increasing the field of view (i.e. the width of the image) increases the number of lines of sight in each image (L) and therefore reduces the frame rate.

When write zoom is used, the machine is only scanning a fraction of the anatomy that the full image would have shown, and so it needs fewer lines of sight. This increases the frame rate.

Multiple focus, compound imaging, harmonics and colour Doppler all increase the number of transmit pulses required for each line of sight, and so they reduce the frame rate.

Chapter 11: Bioeffects and safety

1. A bioeffect is any physical or chemical change induced in the body due to ultrasound exposure.
    A biohazard is a bioeffect which is potentially harmful to the tissues.

2. Thermal: heating of tissues due to the absorption of ultrasound by the tissues.
    Mechanical: cavitation, where small gas bubbles in the body oscillate due to the pressure variations in the ultrasound wave.

3. Thermal Index: calculated as

    $$TI = \frac{W}{W_{deg}}$$

    where W is the transmitted ultrasound power and W_{deg} is the machine's estimate of the power level that would give rise to a temperature rise of 1°C in the patient's tissues.
    Mechanical index: calculated as

    $$MI = \text{constant} \times \frac{P_r}{\sqrt{f}}$$

    where P_r is the maximum rarefaction pressure of the transmitted ultrasound pulse and f is the frequency. This indicates the likelihood of cavitation occurring if gas bubbles are present in the tissues under examination.
    Ultrasound equipment manufacturers are required to display these indices so the user is aware of the exposure level they are using. Modern guidelines allow higher levels of transmitted ultrasound power than in the past, so the user has an obligation to ensure that the machine is operating safely.
    The primary safety guideline is the ALARA Principle, which states that the patient's exposure to ultrasound (both the intensity and duration) should be 'as low as reasonably achievable'.
    Organisations such as the British Medical Ultrasound Society, the American Society for Ultrasound in Medicine and the World Federation for Ultrasound in Medicine have published detailed guidelines recommending maximum TI and MI values for various clinical application areas.

4. The Thermal Index is based on a mathematical model of the way tissues heat, as well as the machine's knowledge of the ultrasound exposure parameters. Different models are used depending on whether the region around focus contains only soft tissue (TIs) or also contains bone (TIb). A third model is used for transcranial ultrasound where bone is adjacent to the probe (TIc).

5. Temporal peak intensity: instantaneous maximum value during the transmit pulse.
Pulse average intensity: average intensity during the transmit pulse.
Temporal average intensity: intensity averaged over one or more complete transmit/receive cycles.

6. The duty factor is the fraction of time for which the machine is transmitting ultrasound. It can be calculated as

$$DF = \frac{\tau}{PRP}$$

where DF is the duty factor, τ is the transmit pulse duration and PRP is the pulse repetition period (the reciprocal of the PRF). The temporal average intensity is simply the pulse average intensity multiplied by the duty factor.

7. Diagnostic ultrasound has a good safety record.
Current ultrasound equipment can expose the patient to relatively high intensities, particularly in pulsed Doppler mode. Users therefore have a responsibility to use ultrasound only when clinically indicated (or for research or educational purposes).
The ALARA principle should always be followed (keep exposure levels and time As Low As Reasonably Achievable).

Chapter 12: Additional modes and capabilities

1. Several images are acquired, with the beams steered in a different direction for each image. The images are combined to produce a single compound image. The result is more complete display of tissue boundaries, smoother speckle and reduction of a number of image artifacts (such as shadowing and enhancement). The major disadvantage is that the machine must acquire several images to form each compound image, so the effective frame rate is substantially reduced; also, some of the artifacts that are reduced (e.g. shadowing) can be useful diagnostic signs.

2. In areas where the transmit pulse intensity is relatively high, the transmitted ultrasound pulse becomes progressively distorted as it travels through the patient's tissues. Specifically, the positive pressure peaks travel slightly more quickly and the negative pressure peaks travel slightly more slowly than the usual propagation speed.
    As a result, the transmit pulse contains some energy at harmonic frequencies (multiples of the original transmit frequency). Any echoes caused by the distorted transmit pulse will themselves be distorted and so they will also contain energy at harmonic frequencies. Echoes from regions where the transmitted ultrasound intensity is lower will not be distorted so they will not contain energy at harmonic frequencies.

3. In harmonic imaging the machine removes the transmitted frequency (fundamental frequency) components of the received echoes, leaving only the energy at the harmonic frequencies (mainly the second harmonic). The image therefore displays only echoes that come from tissue regions where the transmitted ultrasound is significantly distorted.
    So the image only contains echoes from the central part of the main ultrasound beam, where the intensity is relatively high. This dramatically reduces beam width, sidelobe and slice thickness artifacts.
    Since the build-up of the transmit pulse distortion is progressive, the transmit pulse is relatively undistorted in the

first centimetre or so of its travel and hence reverberation artifact from superficial tissues is also significantly reduced.

4. When ultrasound is scattered by small gas bubbles (such as those in ultrasound contrast agents) the echoes are distorted and so they contain harmonic energy. This occurs because the bubbles vibrate non-linearly and resonate in the oscillating pressure field created by the ultrasound.
    Harmonic imaging enhances the visibility of the echoes from the contrast bubbles, since it only shows distorted echoes.

5. Harmonic imaging requires two transmit pulses for each line of sight and so it halves the frame rate compared with standard imaging.
    As the echoes travel through the patient's tissues, the harmonic energy is attenuated far more rapidly than the fundamental frequency component since the frequency is higher. So the depth of penetration is reduced in harmonic imaging.

6. Ultrasound contrast agents consist of small gas bubbles which travel within the patient's blood, greatly enhancing its echogenicity. The bubbles are small ($<5\mu$) and they are encapsulated with a thin shell to slow their absorption in the blood. They must also be non-toxic.
    Clinical applications include:
    Improved heart wall outlining.
    Improved Doppler signal quality in difficult areas such as transcranial Doppler.
    Imaging of the liver following contrast administration to detect regions with abnormal function.

7. The machine sweeps the scan plane through a volume of tissue, storing the acquired images. It then combines these images to produce one or more visualisations (or renderings) of the 3D ultrasound echo data.
    The machine achieves the volume scan by (a) mechanically rocking the transducer backwards and forwards using a small motor enclosed in the probe, or (b) electronically steering the scan plane through a range of angles using a matrix array transducer.

8. Surface rendering.
    Orthogonal display.

Reslicing.
Maximum projection rendering.

9. The machine selects a small region of interest (ROI) in the image. It then compares the speckle pattern within this ROI over successive images and detects any movement. From this it can measure both the speed and direction of tissue motion.

10. Tissue can expand and contract, either as part of its function (e.g. the myocardium) or in response to an applied force. The fractional change of a dimension (such as length or volume) is called strain.
The machine can measure strain by defining a number of ROIs in the tissue and comparing their movement. This allows the machine to calculate relative motion, for example, shortening or lengthening of the tissue.

11. Ultrasound elastography assesses the stiffness of tissues by observing how they move in response to a mechanical force. The tissues may be manually compressed (e.g. by pressing on the probe).
Alternatively, a shear wave may be generated by the machine and its progress through the tissues observed, yielding a measure of stiffness.
Many machines provide only a qualitative map showing increased and decreased stiffness. Others provide quantitative information. Promising application areas include the liver and breast.

12. The term synthetic aperture imaging refers to a new way of generating ultrasound images where large amounts of echo data are collected from each transmit pulse. From this data the machine may calculate multiple lines of sight or even an entire image. Advantages of this method include:
A dramatic increase in frame rate, allowing the use of new techniques such as shear wave elastography and vector flow imaging.
Optimal focus throughout the image, with no need for the user to select the depth of focus.

13. The term Artificial Intelligence (AI) refers to the use of software to carry out some tasks that would normally require human intervention. Examples include:

Automated rendering in 3D fetal imaging.

Automated recognition of anatomical landmarks to facilitate measurements, for example in the heart.

14. The probe is moved over the patient's skin in line with the scan plane. As each image is acquired the machine compares it with the stored image and adds any new information to the stored image so that it is progressively extended.

    It is vital that the scan plane is consistently maintained as the probe is moved.

www.ingramcontent.com/pod-product-compliance
Lightning Source LLC
Chambersburg PA
CBHW061840220326
41599CB00027B/5357